# 3rd Grade Science Volume 3

© 2013 Todd Deluca
OnBoard Academics, Inc
Newburyport, MA 01950

800-596-3175
www.onboardacademics.com

# Table of Contents

**Using Senses to Survive**                                    **3**

    **Using Senses to Survive Quiz**                           **12**

**Ways an Object will Move**                                   **13**

    **Ways an Object will Move Quiz**                          **19**

**Forces and Motion**                                          **20**

    **Forces and Motion Quiz**                                 **26**

**Magnets**                                                    **27**

    **Magnets Quiz**                                           **37**

**Static Electricity**                                         **39**

    **Static Electricity**                                     **44**

# Using Senses to Survive

What are the five senses?

Unscramble the letters to find the answers.

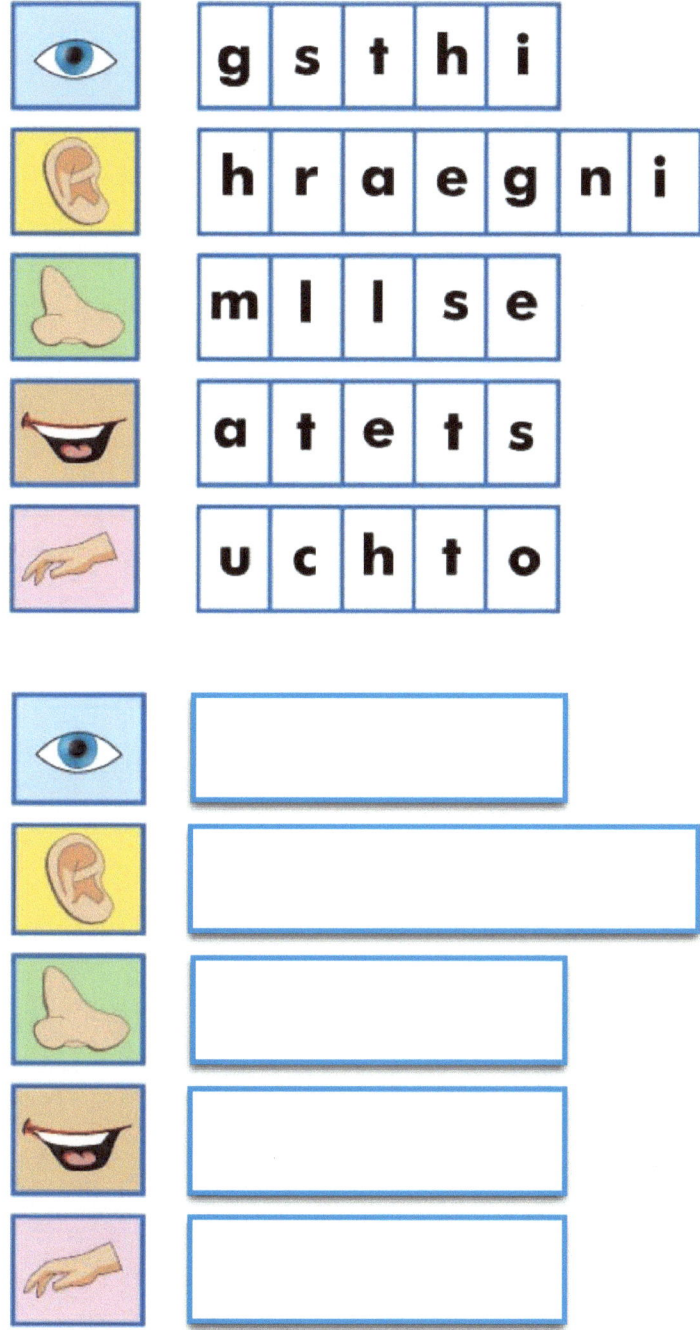

What are the five sense organs.

Label the organs with the suggestions provided.

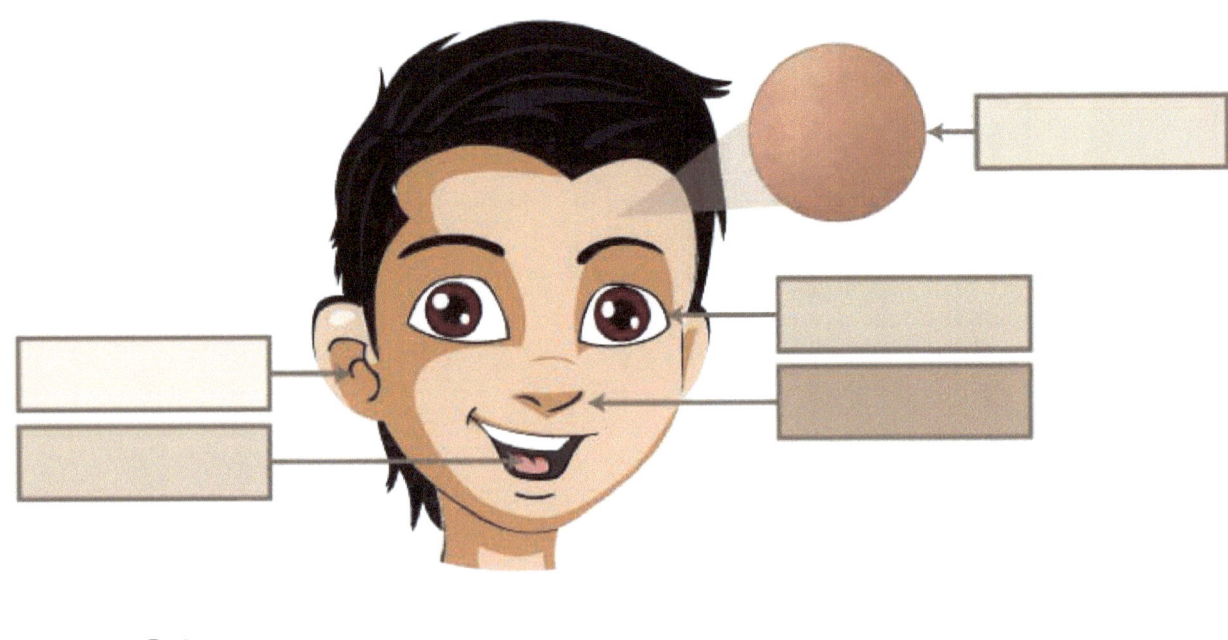

skin    eyes    nose    tongue    ears

**Humans have five sense organs: eyes, ears, nose, tongue, and skin. These organs send messages to the brain which give us the sensation of sight, sound, smell, taste, and touch.**

Connect the expression with the sense(s) being used.

Owww, that's hot!

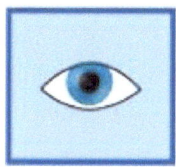

Yuck! The milk is sour.

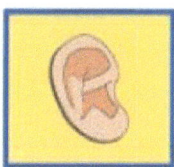

I tripped and fell because it was dark.

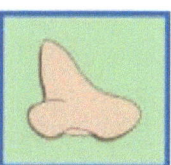

I knew the stove was on fire even before the smoke alarm sounded.

"Roar" Ah-oh we better get going!

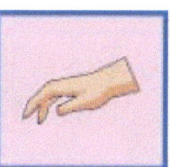

The sun is going down, we should find shelter.

Senses in the animal kingdom

Which senses are these animals noted for?   Connect the animal with their sense specialty.

**smell**     **sight**     **touch**     **taste**     **hearing**

Using Senses to Survive

 Eagles use their keen sense of sight to spot prey. They can see fish in the water, even thought the water and fish are similar shades, from hundreds of feet above. They can spot a rabbit from a mile away.

 Using taste buds located on the tongue humans have a very good sense of taste. This helps to identify things that taste good from those that might be harmful. In addition to the tongue, did you know that much of our sense of taste comes from our sense of smell?

 A cat has an amazing sense of touch using its whiskers, paws and nose. It also has touch receptors on the hairs of its body. This helps it to identify dangerous situations.

 Although dolphins have very good eyesight, its not easy to see under water and so they use sound to find prey and to communicate. The dolphins auditory nerves are twice the size of humans. They can hear sounds from far away and sounds that we can't hear at all.

 Bears have an exceptional sense of smell. A bear can smell a dead animal from 20 miles away. That's the reason you should keep food outside of your tent when camping in the forest.

Some animals use their senses differently than we do.

Can you connect the animal with the description of the way they use their senses?

I have my eyes and ears close when hunting for food. I use my sense of smell to dig in the mud for worms.

My hearing organs are called tympani and are located in my front legs.

I don't have any eyes but I do have organs that can detect light and I have taste receptors at the front of my body.

I use my very sensitive tongue to smell and find prey.

Some Animals use other special senses.

Many animals including humans use some or all of the five senses including sight, hearing, smell taste and touch in order to find prey, avoid danger and find mates.

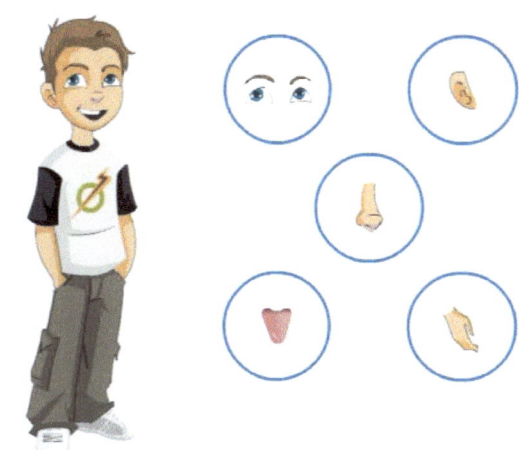

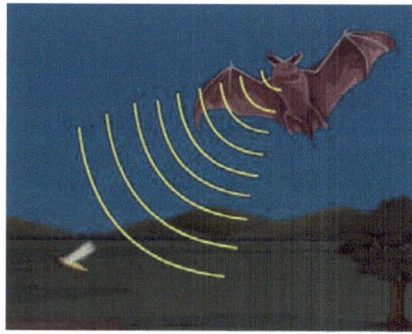
Some animals uses senses that we don't have in order to survive. Bats send out ultrasonic or very high frequency squeaks to avoid obstacles and find prey. They then calculate the time it takes for the squeak to bounce back to them and can locate the prey or avoid the obstacles. This special sense is called echo location.

Other animals like sharks have an electric sense which enables them to track prey using the weak electric fields produced by other fish. Electric eels go one better and use an electric shock to stun animals so they can capture and eat them.

Rattlesnakes have organs that can sense heat so that they can see and hunt warm blooded prey in the dark. We call this sense infra red vision.

Other animals such as sea turtles and birds are believed to sense the Earth's magnetic field that helps them to stay on course during lengthy migrations.

Which special sense does each animal use.  Label each box with the special sense.

echolocation

infrared vision

magnetic sense

electric sense

## Using Senses to Survive Quiz

1.  How many sensory organs do humans have? _____

2.  If you close your _____, you cannot see.
    a. ears
    b. eyes
    c. skin
    d. nose

3.  What organs do humans use to taste food?
    a. _____
    b. _____

4.  You recognize rotting food by smelling.  True or false?

6.  What sense does an eagle use to find prey? _____

7.  Bears are noted for their keen sense of _____.

# Ways an Object will Move

Label each item with the way in which it will move.

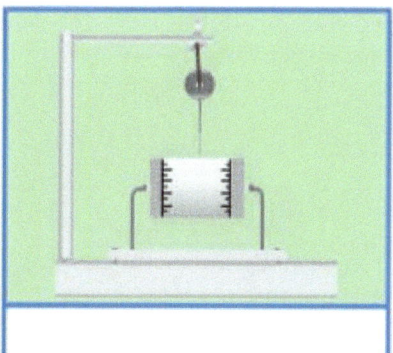

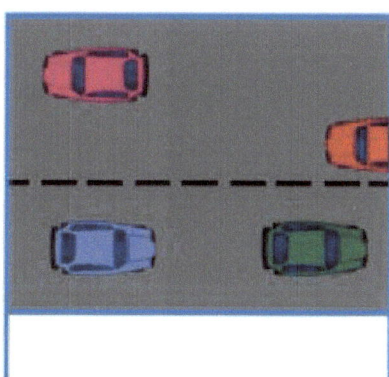

**round and round**     **straight line**     **side to side**

**back and forth**      **zig zag**          **up and down**

Pushes and Pulls Change Motion

Label each action push or pull.

**push**

**pull**

A push or a pull is a force, and forces can change the way that things move. For example, a push or a pull can make an object start moving or make a moving object change its direction or slow down.

How are these objects moving?

Read the descriptions and then label each item by the way in which it moves.

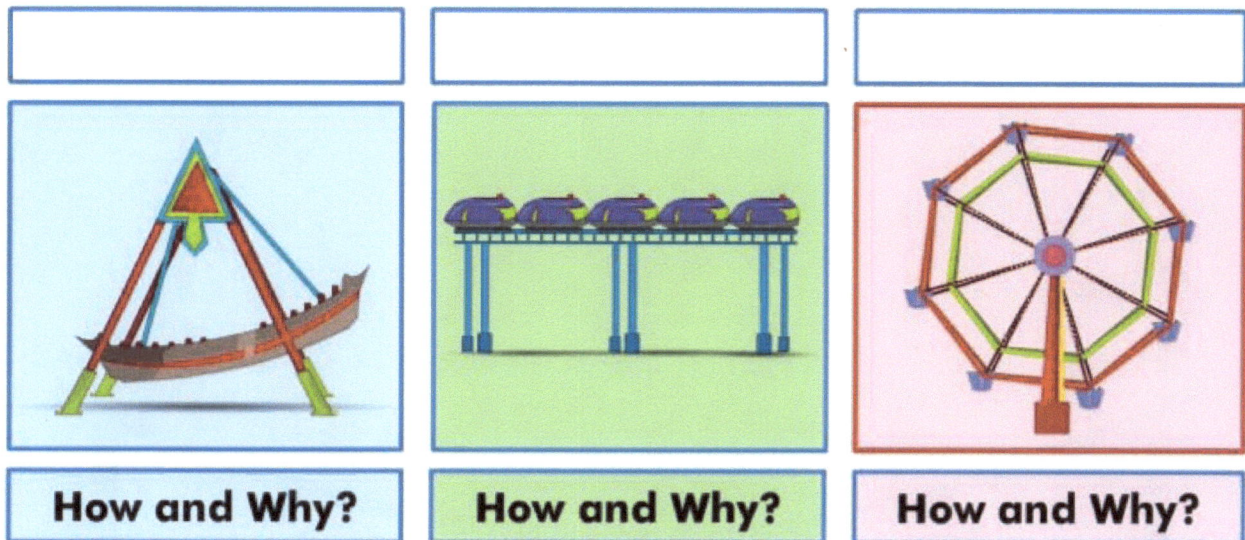

| | | |
|---|---|---|
| **How and Why?** | **How and Why?** | **How and Why?** |

**Motion that goes round and round happens because the middle point is fixed. In this Ferris wheel ride, the force comes from the center.**

**Motion that goes in a straight line happens because there is a force that moves it in one direction.**

**Motion that goes back and forth (or side to side) happens because there is a force that moves it in one direction and a force that moves in back in the opposite direction.**

Will this ball ever stop rolling? Why? _____

_____

_____

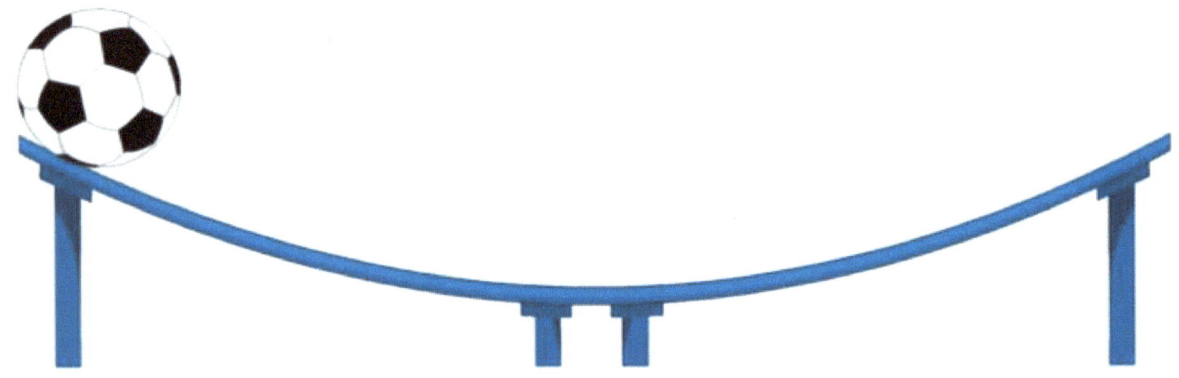

**Friction is a force that slows down moving objects. If the ball was rolling on a bumpy surface, it would slow down more quickly because there would be more friction. If the ball was rolling on a smooth surface, it would take longer to slow down because there would be less friction.**

Which way will the ball move and by how much?

Use the red arrows to indicate the direction and the amount the know will move.  Please note that some red arrows are longer than others indicating more movement.

Name: _____

## Ways an Object will Move Quiz

1. _____ is a result of force.
   a. Velocity
   b. Motion
   c. Revolution
   d. Tension

2. The motion that goes _____ happens because the middle point is fixed.
   a. back and forth
   b. round and round
   c. up and down
   d. side to side

3. Motion that goes _____ happens because there is a force that moves the object in one direction.
   a. up and down
   b. side to side
   c. in a straight line
   d. in a zig zag pattern

4. Friction is a force that pushes in the direction that an object is moving, causing the motion to speed up. True or false?

# Forces and Motion

A force is a push or a pull.

Place a √ in the box to indicate if the motion is a push or a pull.

| | push | pull |
|---|---|---|
| kick | | |
| slide | | |
| drop | | |
| throw | | |
| roll | | |

Roll

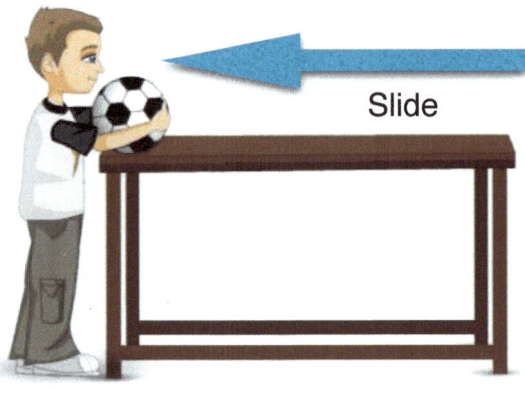

Slide

**The more massive an object is, the greater the force that is needed to make it move.**

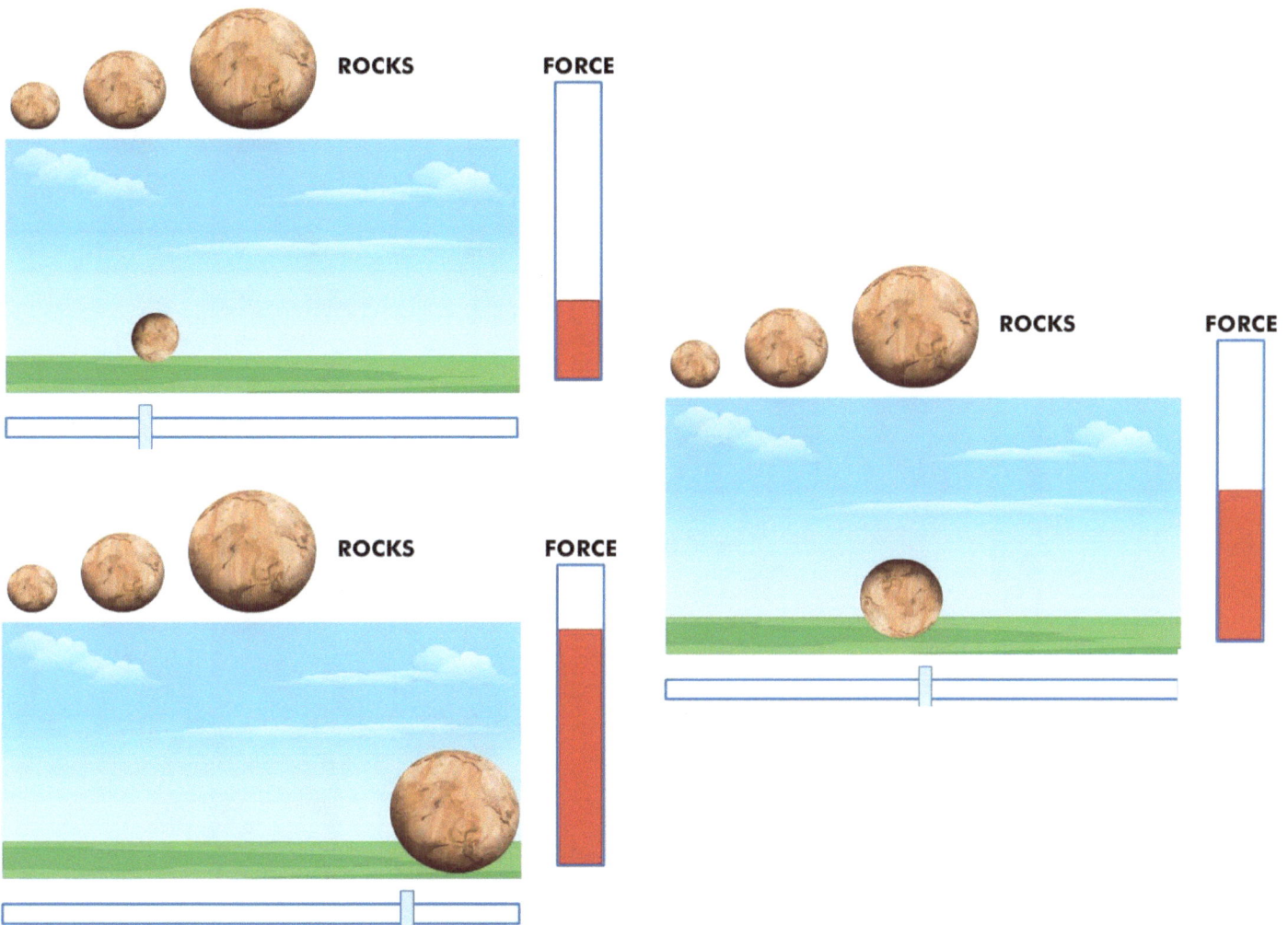

ROCKS    FORCE

ROCKS    FORCE

ROCKS    FORCE

**A larger force will move an object farther and faster than a smaller force. With the same amount of force, a smaller, lighter object will move farther and faster than a heavier one.**

Which force and mass combinations will make these outcomes.

Either draw the answer in the box provided or connect the empty box with the correct illustration.

Force and Motion Review

Connect the vocabulary word with the correct definition.

| | |
|---|---|
| **mass** | The location of an object. |
| **force** | The change in an objects position. |
| **motion** | A push or a pull that can change an objects motion. |
| **speed** | A measure of how fast an object moves. |
| **friction** | A force that slows the motion of an object. |
| **position** | If this is greater, more force is need to make this object move. |

We don't always notice motion.

Some things move so quickly we don't notice that they are moving. Light moves at about 300,000 km per second. That's fast enough to circle the Earth seven times in one second.

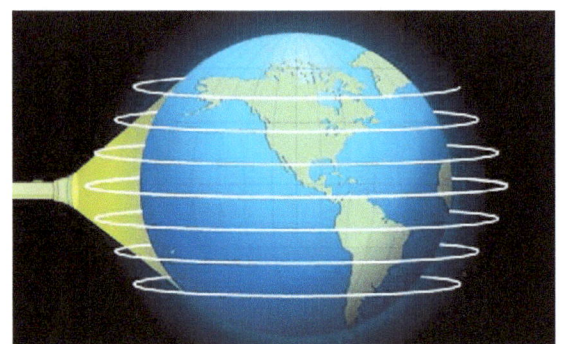

The Earth also travels very quickly. At the equator, the Earth is spinning about 1,700 km per hours. But the rotational speed seems like a snail's pace when you consider that the Earth orbits the sun at about 107,000 km per hour.

If we are rotating this quickly and traveling around the sun so fast, why don't we notice?

We don't notice because everything around us is moving at exactly the same speed. Think about flying in an airplane. You don't realize you are moving that fast because everything around you including the plane is also moving at exactly the same speed.

On the other hand, some things are moving so slowly you don't notice them moving either. Earth's continents are moving at 2.5 centimeters per year. That might not sound like much but in a million years that's 25 km and a million years isn't very long in geological terms.

## Forces and Motion Quiz

1. _____ always has a direction, and is caused by force.
   a. Work
   b. Energy
   c. Motion
   d. Velocity

2. _____ is a force that slows down moving objects.
   a. Acceleration
   b. Motion
   c. Velocity
   d. Friction

3. The _____ an object, the _____ the force required to move the object.
   a. lighter, greater
   b. heavier, lesser
   c. heavier, greater

4. When a large force is applied, an object moves a _____.
   a. smaller distance
   b. larger distance

# Magnets

James dropped a paper clip and a key into a drain. James had the bright idea to retrieve them using a magnet. James is a thinker! However, James retrieved the paper clip but the magnet was not attracted to the metal key.

Do you know why?_____

**Magnets are objects that attract magnetic materials. The four most common magnetic materials are iron, nickel, cobalt, and steel (which is a mixture of iron and other materials). James can use the magnet to retrieve his paper clip, which is made of steel, but the key is made mostly of brass, which is not magnetic.**

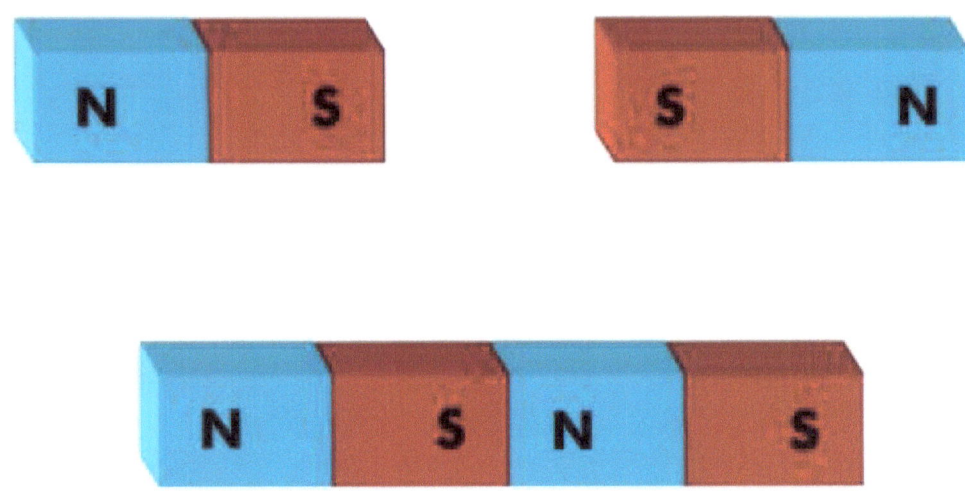

A magnet has two poles—one at each end—called a north pole and a south pole. **Opposite**, or unlike, poles **attract** (pull towards each other), while **similar**, or like poles, **repel** (push away from) each other.

Will these poles attract or repel?

Place an **A** for attract.
Place an **R** for repel.

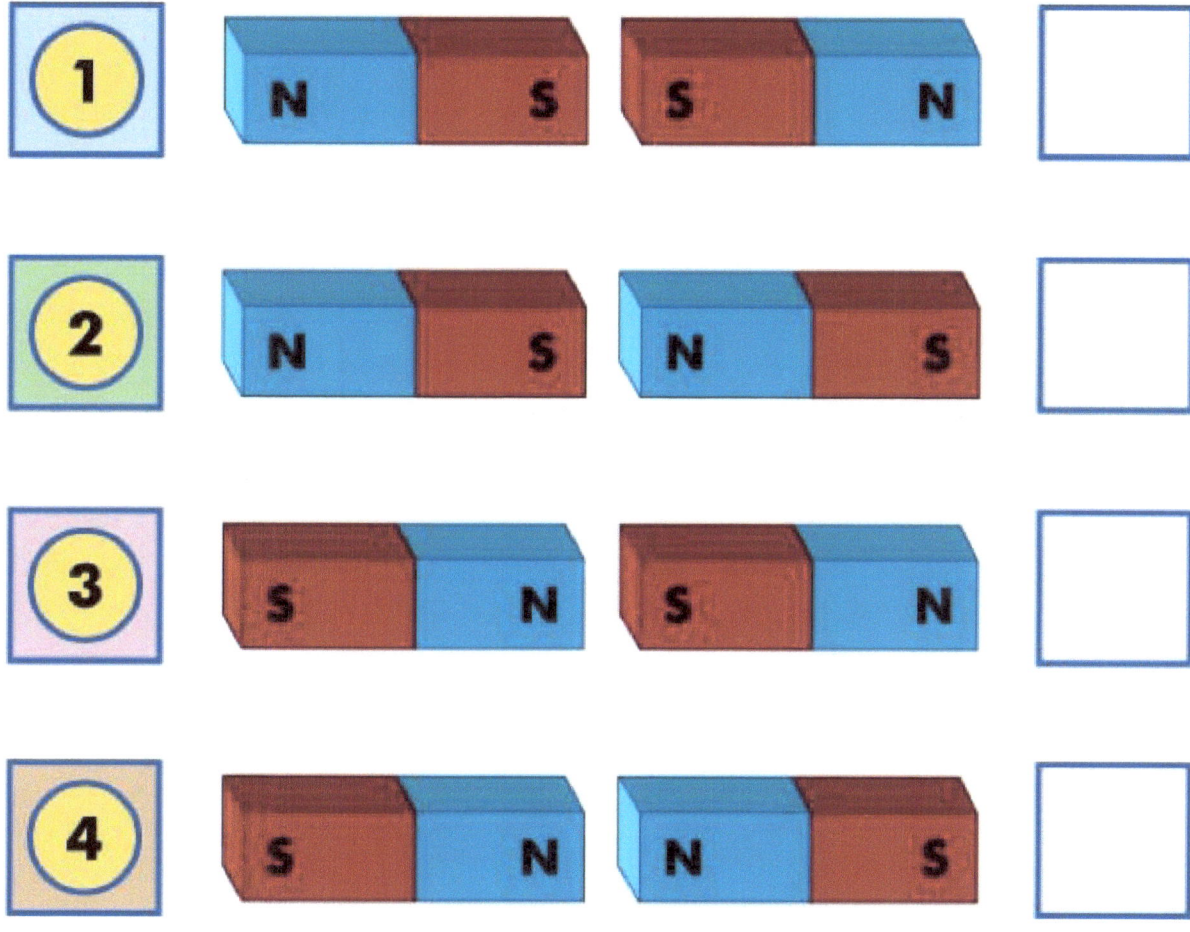

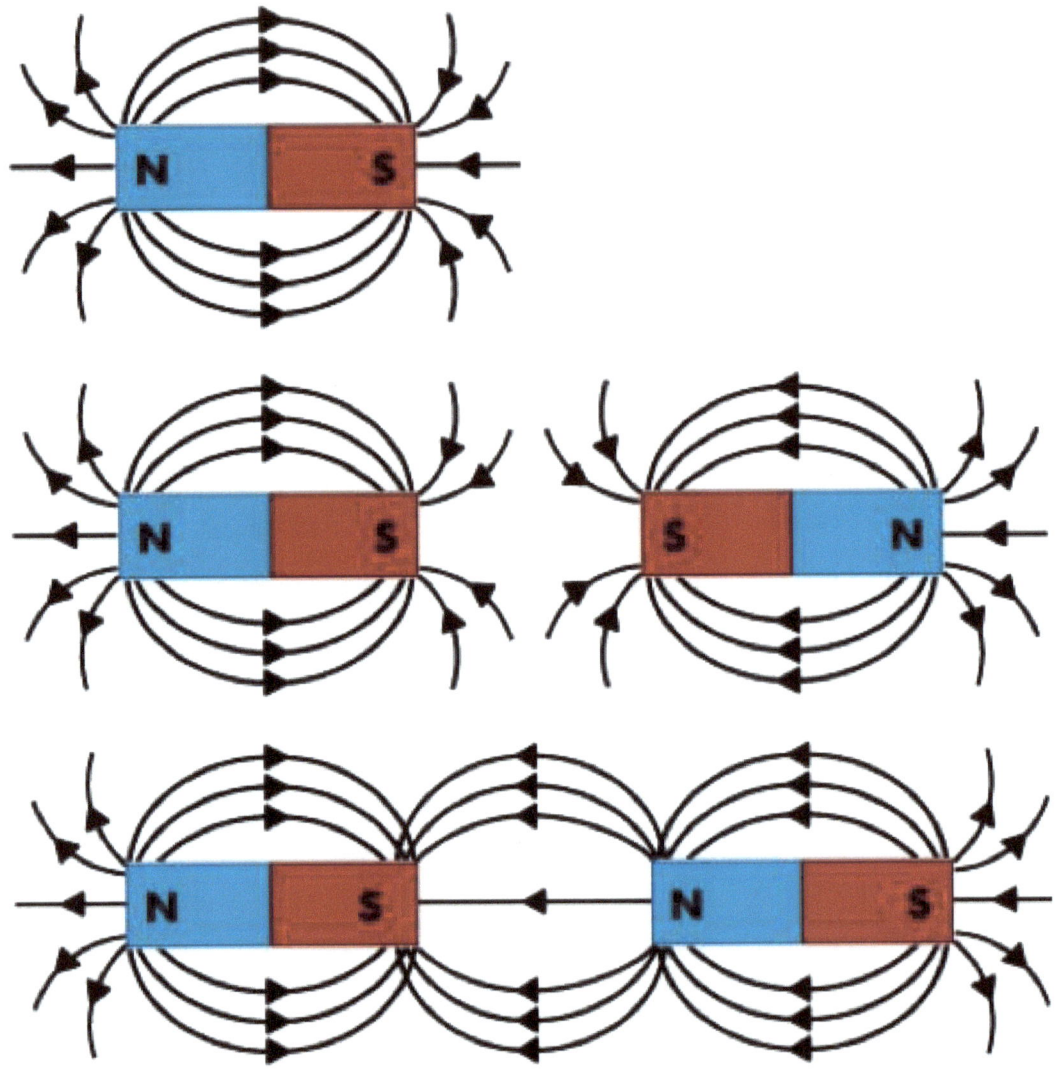

The area around a magnet where the force acts is called the  magnetic field  . The forces in a magnetic field can be demonstrated by sprinkling iron filings around a magnet. The filings show the forces of attraction and repulsion around a magnet which is strongest at the poles.

Circle the items that the magnet will pick up.

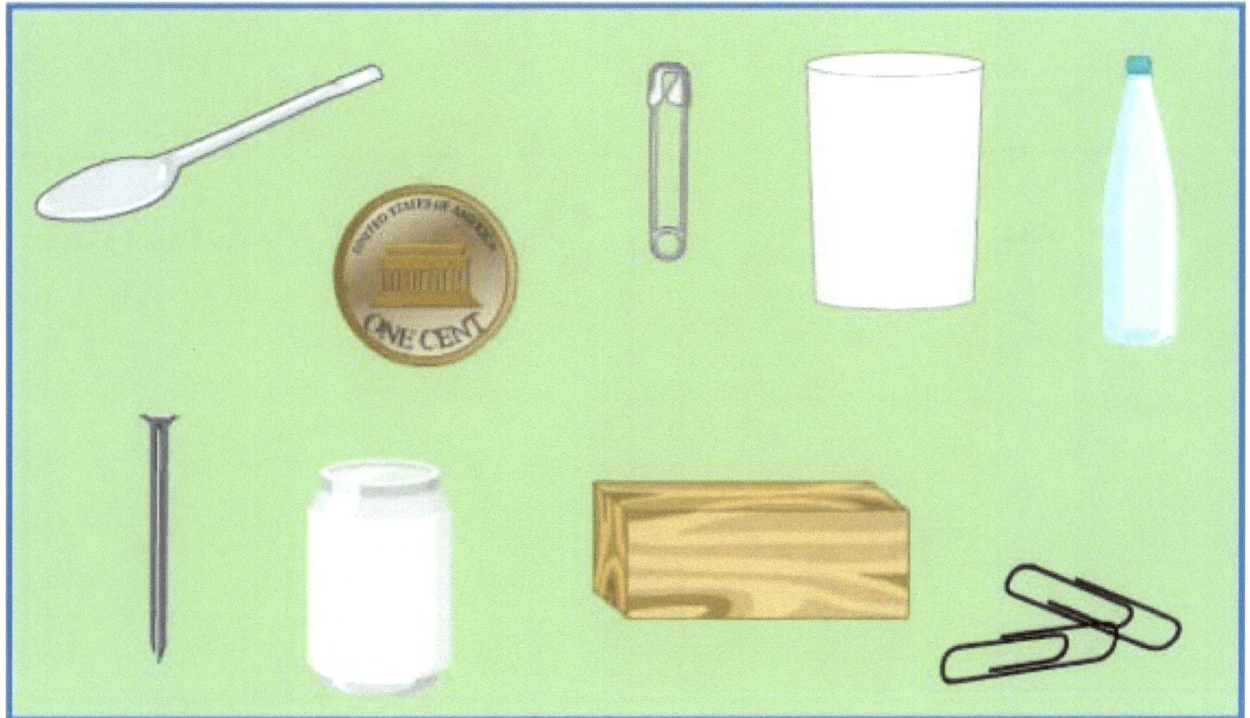

**Metals such as iron, nickel, and cobalt all have particles that react to a magnet's field and so we call these metals magnetic. The particles in other metals such as aluminum and copper don't react to a magnet's field and so these metals aren't magnetic. A penny is made up mostly of zinc which isn't magnetic.**

Sort these materials by magnetic or non magnetic.

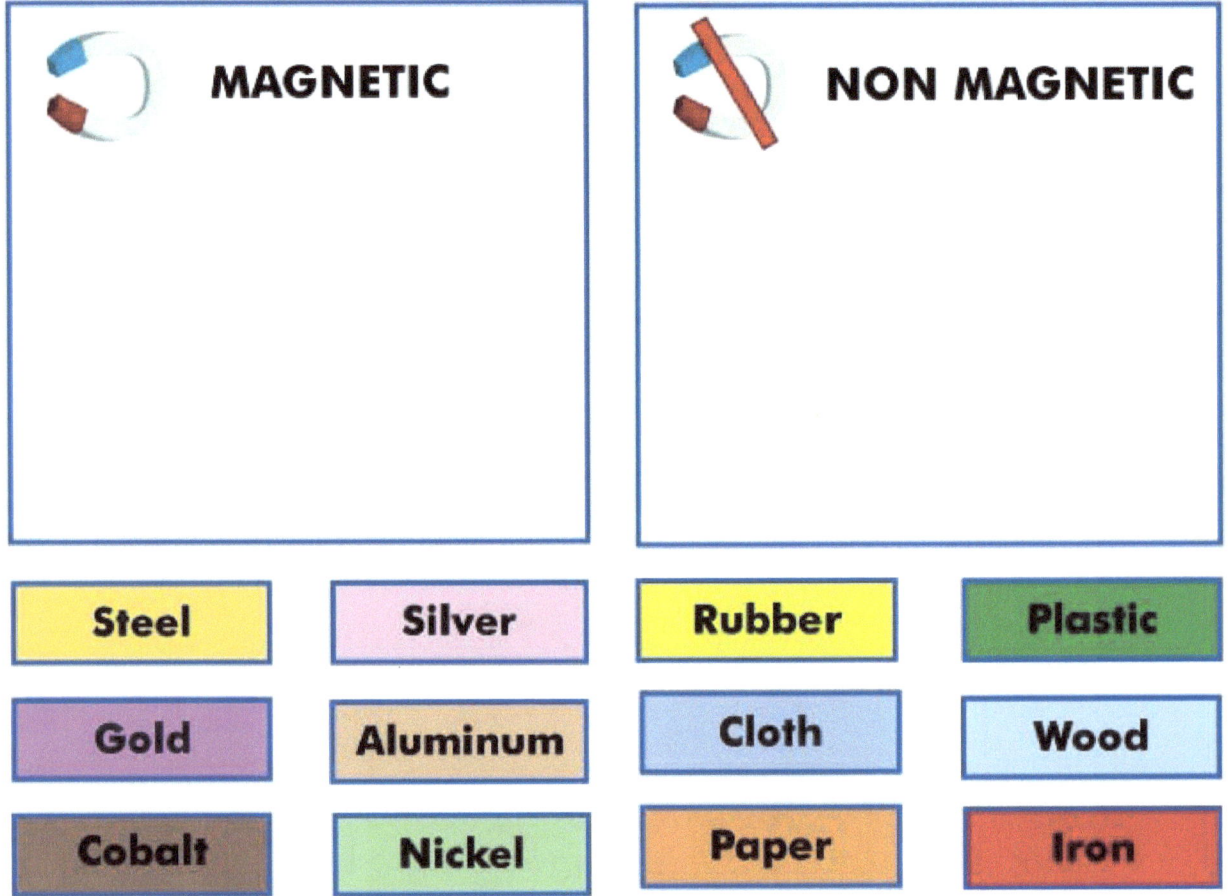

| MAGNETIC | NON MAGNETIC |

| Steel | Silver | Rubber | Plastic |
| Gold | Aluminum | Cloth | Wood |
| Cobalt | Nickel | Paper | Iron |

www.onboardacademics.com

The world's largest magnet.

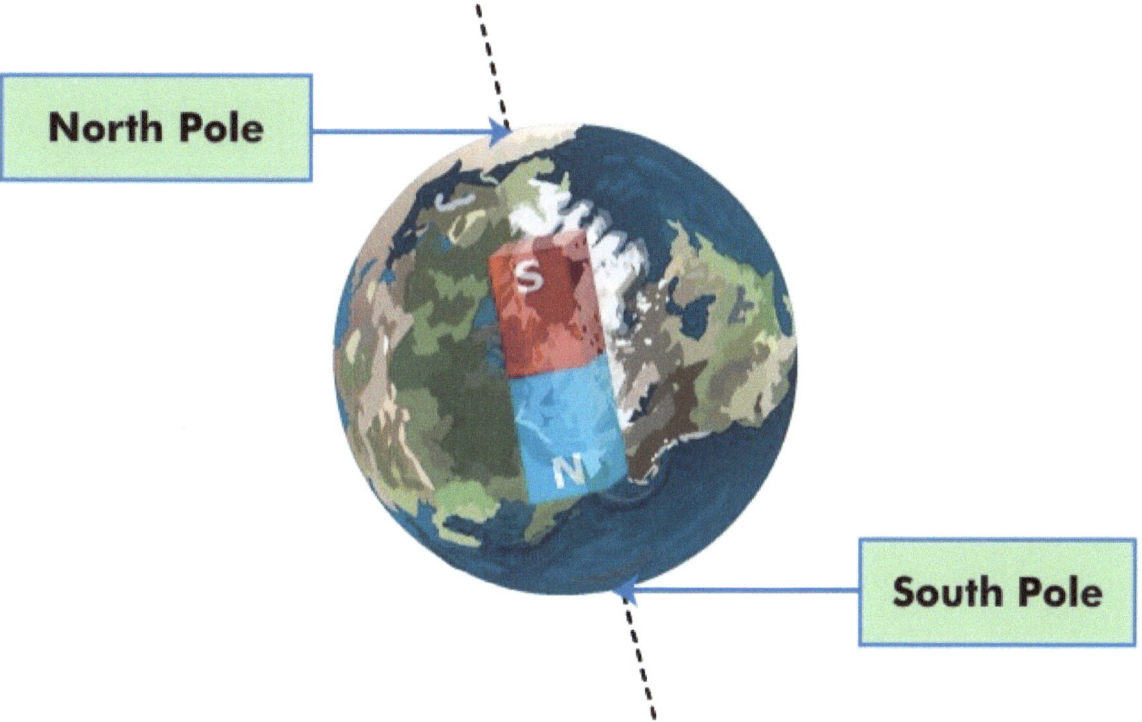

**North Pole**

**South Pole**

There is a large magnetic field surrounding the Earth. This is believed to be caused by the hot metals in Earth's core. As a result, the Earth itself acts like a magnet with two poles. A little confusingly, Earth's magnetic south pole is located somewhere close to Earth's geographic North Pole, while Earth's magnetic north pole is located close to Earth's geographic South Pole.

A compass uses a magnet that is shaped like a needle to give us directions. Because the end of the needle is actually the north-seeking pole of a magnet, it is always attracted to the Earth's magnetic south pole, which, as we've learned, is Earth's geographic North Pole. Once we know which direction north is, it's easy to identify south, west, and east.

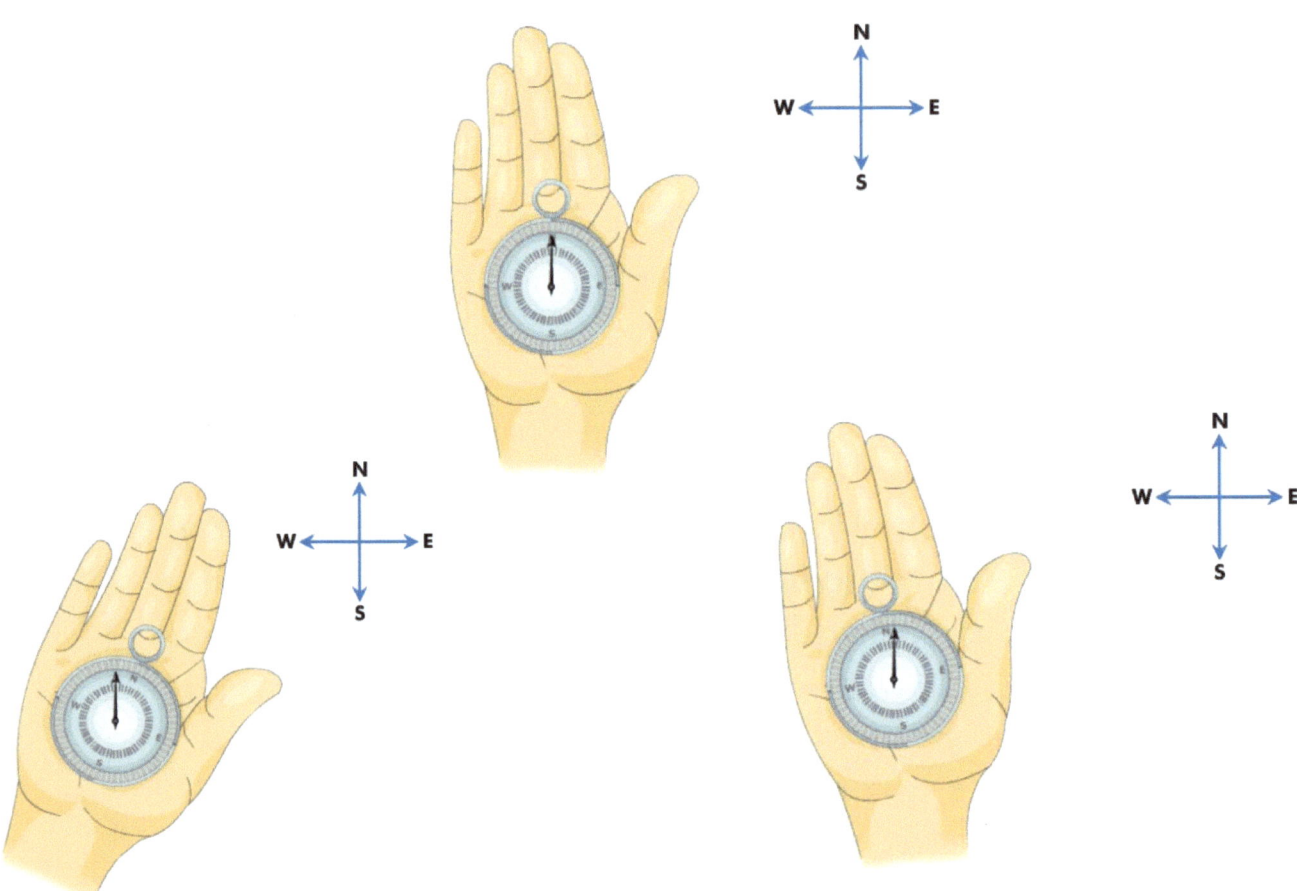

Apart from compasses where else are magnets found?

Place a √ for magnet
Place an X for no magnet

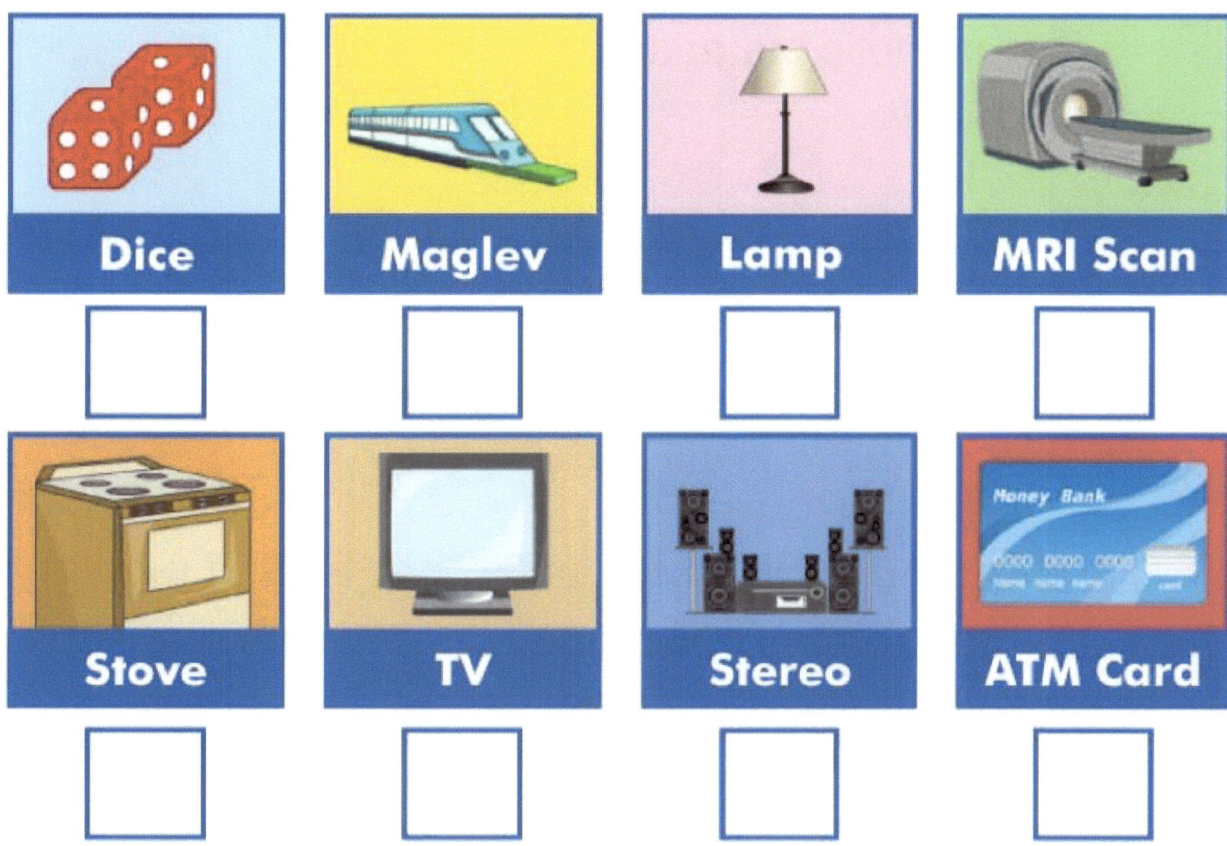

Due to their ability to hold, control, and separate materials, magnets are used in many appliances including generators, telephones, televisions, and computers. Magnetism is also the force at work in Maglev trains, some of the fastest trains in the world.

## Magnets Quiz

1. Any magnet has two ends called:
   - a. magnetism
   - b. magnetic material
   - c. poles

2. The Earth's magnetic field is believed to be caused by the hot liquids in its core.  True or false?

3. The like poles of two magnets attract each other.  True or false?

4. The area around a magnet where the force of the magnetism can be experienced is called the _____.
   - a. south pole
   - b. north poe
   - c. magnetic field
   - d. magnetic personality

5. The magnetic force is stronger in the middle than at the poles.  True or false?

6. A compass uses a magnet shaped like a needle to tell time.  True or false?

Newburyport, MA 01950

1-800-596-3175

OnBoard Academics employs teachers to make lessons for teachers! We create and publish a wide range of aligned lessons in math, science and ELA for use on most EdTech devices including whiteboard, tablets, computers and pdfs for printing.

All of our lessons are aligned to the common core, the Next Generation Science Standards and all state standards.

If you like our products please visit our website for information on individual lessons, teachers licenses, building licenses, district licenses and subscriptions.

Thank you for using OnBoard Academic products.

# Static Electricity

Static Electricity

Meet Owen.  Owen is in a room alone with a balloon. Owen decides to rub the balloon agains his hair.

Two things happen.  Owen's hair stands up and is strangely attracted to the balloon and the balloon sticks to the wall like magic.

What's going on here?

**The boy's hair stands on end and the balloon sticks to the wall because of static electricity. To understand static electricity, we need to learn a little bit about like and unlike charges, and atoms: the tiny particles that make up all of the matter around us.**

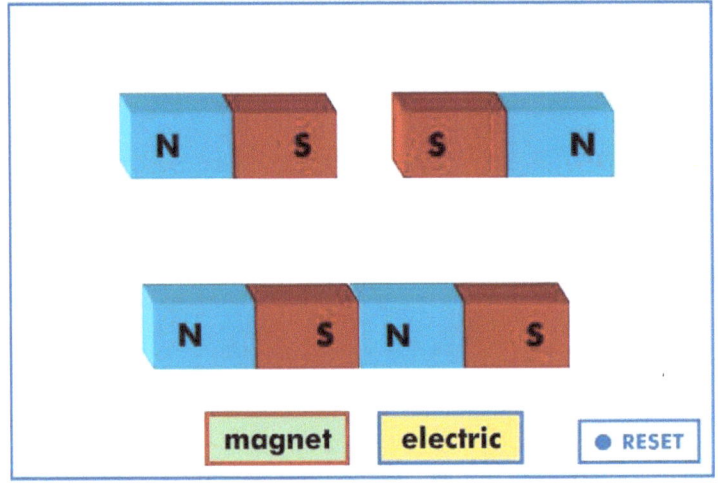

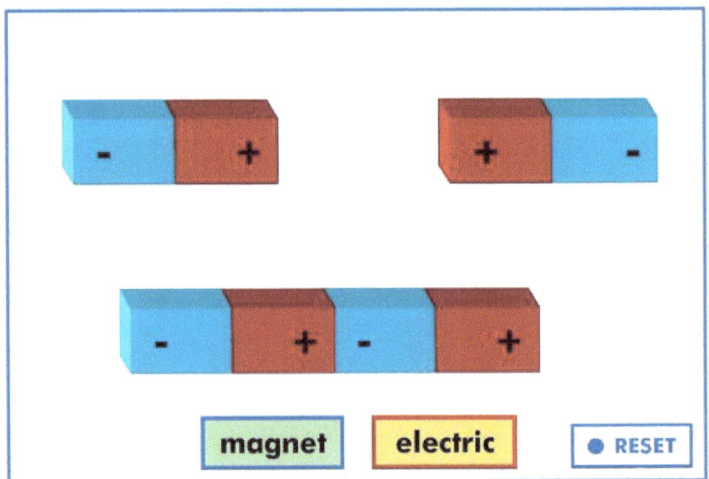

Just as unlike poles attract each other and like poles repel each other, unlike electrical charges attract each other and like electrical charges repel each other.

## Static Electricity

Everything in the universe is made up of tiny particles called atoms. Atoms, which look a little bit like fuzzy balls are very, very small. If we look inside an atom, we will see that there are even smaller particles inside the atom called protons, neutrons and electrons.

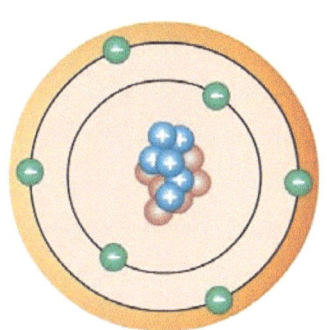

Protons and neutrons are packed together in the center of the atom and form the nucleus of the atom. Electrons zoom around freely outside of the nucleus.

Protons have a positive electrical charge and electrons have a negative electrical charge. Neutrons are neutrally charged.

Ok but what has this to do with static electricity?

Lets go back to the balloon and Owen's bad hair day. When Owen rubbed the balloon against his hair negative electrons from Owen's hair transferred to the balloon.

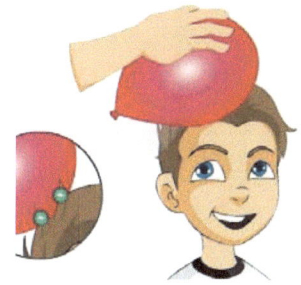

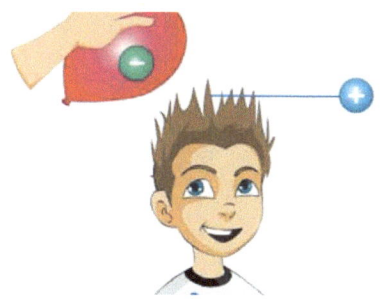

Since Owen's hair now has more positively charged protons, now that the electrons have been transferred to the balloon, we say that Owen's hair is positively charged. Similarly since the balloon acquired all of the negative electrons from Owen it now has more electrons than protons and is negatively charged.

Just like the poles of a magnet, unlike electrical charges attract each other. So when we pull the balloon away from his head the positively charged hair is attracted to the negatively charged balloon and his hair moves toward the balloon.

What about the balloon sticking to the wall? The wall doesn't have any particular charge. Remember that our balloon is negatively charged while the wall is neutrally charged.

However when the negative balloon comes near the neutral wall our negatively charged balloon repels negative electrons on the wall and they retreat. This leaves a positively charge spot on the wall to which the balloon is attracted .

That's static electricity in a nut shell. It occurs because materials gain or lose electrons usually when rubbed together. We call this type of electricity static electricity because the electricity doesn't flow through a conductor such as a copper wire and a circuit.

What is the main principle regarding static electricity? _____

_____

_____

## Static Electricity

1. Static electricity can push and pull objects even if they are not touching.  True or false?

2. When your hair is rubbed with a balloon, _____ rub off your hair and onto the balloon.
   a. electrons
   b. protons

3. Two positive electrical charges _____.
   a. attract
   b. repel

4. The protons and neutrons that are packed together in the center of the atom form the _____.

5. Electrons have a negative charge, protons have a positive charge and neutrons have a _____ charge.

6. Why does Owen's hair lift toward the balloon after Owen rubbed the balloon on his head? _____

_____

_____

Newburyport, MA 01950

1-800-596-3175

OnBoard Academics employs teachers to make lessons for teachers! We create and publish a wide range of aligned lessons in math, science and ELA for use on most EdTech devices including whiteboard, tablets, computers and pdfs for printing.

All of our lessons are aligned to the common core, the Next Generation Science Standards and all state standards.

If you like our products please visit our website for information on individual lessons, teachers licenses, building licenses, district licenses and subscriptions.

Thank you for using OnBoard Academic products.

www.ingramcontent.com/pod-product-compliance
Lightning Source LLC
Chambersburg PA
CBHW050832180526
45159CB00004B/1872